InspireScience

Science Paired Read Aloud

ConnectED.mcgraw-hill.com

STEM McGraw-Hill is committed to providing instructional materials in Science, Technology, Engineering, and Mathematics (STEM) that give all students a solid foundation, one that prepares them for college and careers in the 21st century.

Send all inquiries to:
McGraw-Hill Education
8787 Orion Place
Columbus, OH 43240

ISBN: 978-0-02-136948-5
MHID: 0-02-136948-8

Printed in the United States of America.

1 2 3 4 5 6 7 8 9 EUS 20 19 18 17 16 15

Contents

Vocabulary

ecosystem all the living things in an environment

forest a place where there are many tall trees

habitat a place where plants and animals live

need something you must have in order to live

Iggy Iguana

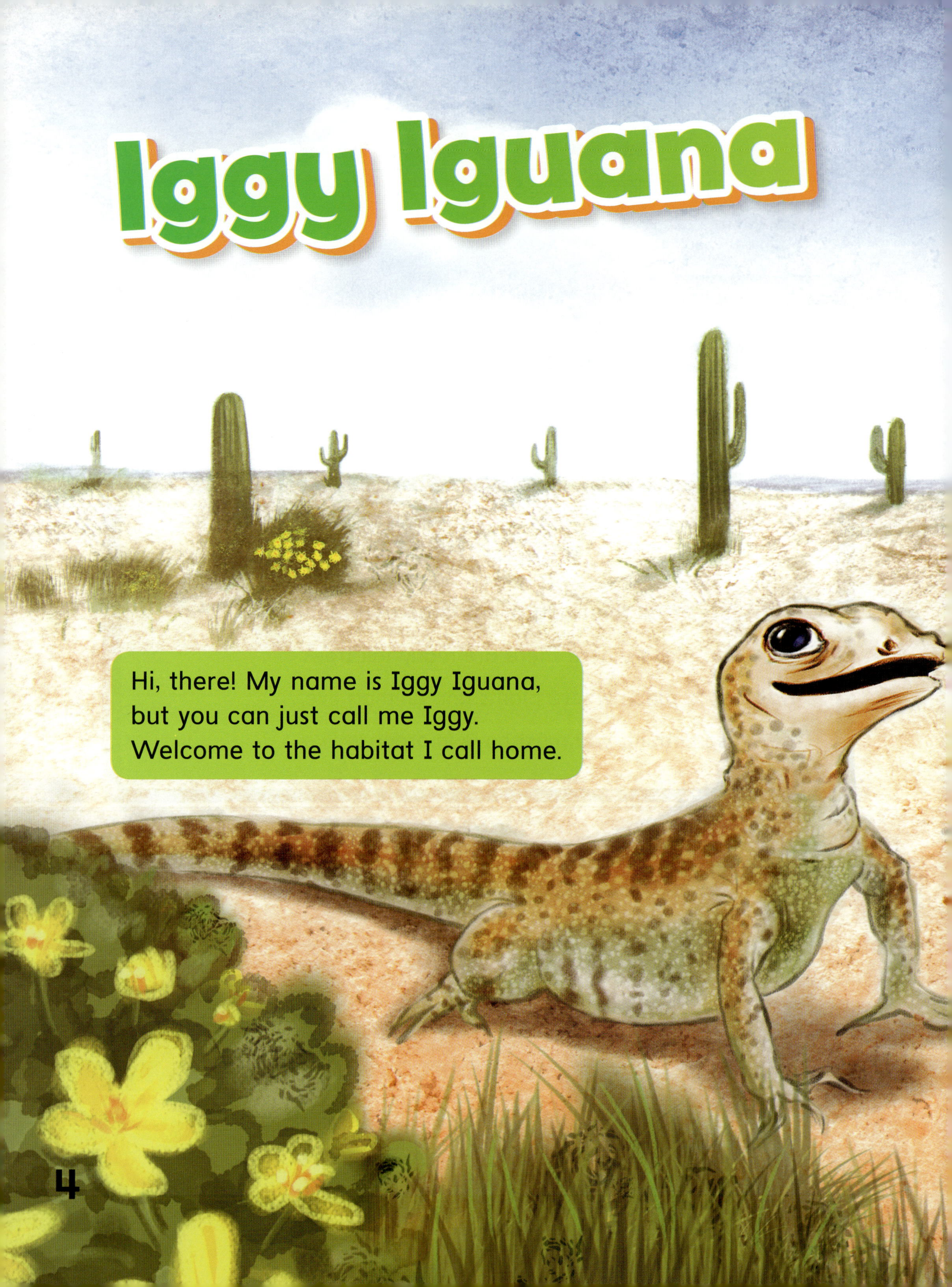
Hi, there! My name is Iggy Iguana, but you can just call me Iggy. Welcome to the habitat I call home.

Let me show you the Sonoran Desert.
I think seeing where someone lives always
tells you a lot about a person—or
an iguana, in my case. Come with me!

I've been living here in the desert for as long as I can remember. It's a dry place, but it suits me well. My body likes to be dry instead of wet.

I love how hot it gets here. The sand feels great on my claws! I like to stay close to the bushes on flat land, and sometimes I like to climb up to the rocky streambeds.

I'm a kind of lizard. My body is short, less than the length of your ruler. I have a very long tail that has brown stripes. My tail is longer than my body! My head is small, and my legs are short. There is a little green on my body, and my belly is a pale color.

Look how well I blend in with the bushes! I feel safe here because it's hard for other animals to find me. Some animals like to eat other animals. But I like to eat plants the most.

These bushes are my favorite food. I really like eating the small yellow flowers that grow on the bush. I climb right into it and help myself! I also eat insects sometimes.

The best part about being an iguana is that I'm fast! I can run so fast across the desert, you might even miss me. I run my fastest when I'm being chased. There are other animals in the desert that would like to have me for dinner.

When a bird or snake tries to get me, I run away and duck into a hiding spot under my favorite bush.

Thanks for visiting my home! I hope to see you again another time.

Animal and Plant Habitats

Animals and Plants Are Living Things

Remember that a living thing is something that grows, changes, and needs food, air, and water to live. People are living things. Animals and plants are also living things.

©NPS photo by George Marler

Living things have different needs. A **need** is something a living thing must have in order to live.

Animals need food and water in order to live and grow.

What Living Things Need

Animals need food. They also need water and air. Plants need these things too. If living things do not have what they need, they can't live and grow.

Plants get what they need
from where they grow.

(t) Exactostock/SuperStock; (b) Don Paulson Protography/Purestock/SuperStock

Animal and Plant Homes

A **habitat** is a place where animals and plants live. Just as you have a home, plants and animals also have a home. There are many kinds of habitats.

Each habitat is special. You will find different kinds of animals and plants in each habitat. Each place has exactly what the animals and plants need in order to live.

The sea is a habitat.

Forest Habitat

Sun bears live in a **forest** habitat. They eat fruit and berries from forest plants. They eat other forest animals, such as lizards and small birds. The forest has what the bear needs to live. The forest also gets the right amount of light and water for plants to grow.

Deer live in the forest.

Bears drink water from forest streams.

Living Things Work Together

In every habitat, there is an ecosystem. An **ecosystem** is a group of living and nonliving things that work together in their habitat.

Water and land are part of an ecosystem.

Animals and plants live near what they need. Living things in an ecosystem need the other animals and plants to stay alive.

Image Source

Discussion Questions

Fiction

- Where is Iggy's habitat?
- Why is the desert a good place for Iggy to live?
- What is the best part of being an iguana according to Iggy?

Nonfiction

- What is a habitat?
- How do plants and animals get what they need from their habitat?
- What kind of habitat do you live in?